1 MONTH OF FREE READING

at
www.ForgottenBooks.com

By purchasing this book you are eligible for one month membership to ForgottenBooks.com, giving you unlimited access to our entire collection of over 1,000,000 titles via our web site and mobile apps.

To claim your free month visit:
www.forgottenbooks.com/free892563

* Offer is valid for 45 days from date of purchase. Terms and conditions apply.

ISBN 978-0-266-80802-2
PIBN 10892563

This book is a reproduction of an important historical work. Forgotten Books uses state-of-the-art technology to digitally reconstruct the work, preserving the original format whilst repairing imperfections present in the aged copy. In rare cases, an imperfection in the original, such as a blemish or missing page, may be replicated in our edition. We do, however, repair the vast majority of imperfections successfully; any imperfections that remain are intentionally left to preserve the state of such historical works.

Forgotten Books is a registered trademark of FB &c Ltd.
Copyright © 2018 FB &c Ltd.
FB &c Ltd, Dalton House, 60 Windsor Avenue, London, SW19 2RR.
Company number 08720141. Registered in England and Wales.

For support please visit www.forgottenbooks.com

GEOLOGICAL SURVEY OF CANADA
ROBERT BELL, M.D., Sc.D. (CANTAB), LL.D., F.R.S.

REPORT

ON

GEOLOGICAL EXPLORATIONS

IN

ATHABASKA, SASKATCHEWAN AND KEEWATIN DISTRICTS

INCLUDING

MOOSE LAKE AND THE ROUTE FROM CUMBER-
LAND LAKE TO THE CHURCHILL RIVER,
AND THE UPPER PARTS OF BURNT-
WOOD AND GRASS RIVERS

BY

D. B. DOWLING, B.A. Sc.

OTTAWA
PRINTED BY S. E. DAWSON, PRINTER TO THE KING'S MOST
EXCELLENT MAJESTY
1902

13—FF.

No. 787.

The EDITH *and* LORNE PIERCE
COLLECTION *of* CANADIANA

Queen's University at Kingston

To Dr. ROBERT BELL,
    Acting Director, Geological Survey of Canada.

SIR,—I have the honour to present the inclosed report on a portion of the eastern part of the District of Saskatchewan and parts of the Districts of Athabaska and Keewatin.

The descriptions of the eastern part of these districts is to be found in the accompanying report by Mr. J. B. Tyrrell, and the part for which descriptions are here given is in a general way the western and southern part of the area shown on the accompanying map.

Mr. Tyrrell's report was accompanied by tracings of his surveys, but in order to make a map which would extend as far west as does the sheet to the south of it describing the geology of north-western Manitoba, additional surveys were undertaken, but the limited time at my disposal in the field (less than three months) did not enable me to visit all the localities which seemed of interest.

To illustrate this report and also that by Mr. Tyrrell, I have compiled the accompanying map on a scale of eight miles to an inch, showing the whole area described in both reports. For much of the information shown on the eastern portion, I was obliged to consult Mr. Tyrrell's note-books, but the geology of the country east of the Nelson river is entirely from Dr. Bell's reports.

                I have the honour to be, Sir,
                    Your obedient servant,

                            D. B. DOWLING.

OTTAWA, February, 1902.

NOTE.—*The bearings throughout this report are given with reference to the true meridian. The variation of the magnetic needle in the vicinity of Lake Winnipeg is 16° E. This is found to increase toward the north-west and is on Kississing river about 19° and on the Churchill river 20° E.*

REPORT

ON

GEOLOGICAL EXPLORATIONS

IN

ATHABASKA, SASKATCHEWAN AND KEEWATIN DISTRICTS

INCLUDING

MOOSE LAKE AND THE ROUTE FROM CUMBERLAND LAKE TO THE CHURCHILL RIVER, AND THE UPPER PARTS OF BURNTWOOD AND GRASS RIVERS.

INTRODUCTION.

Early Maps and Surveys.

The early mapping of the northern part of the area which is shown on the accompanying sheet, is no doubt due to the labours of David Thompson, while in the service of the Hudson's Bay Co. His travels, as recorded in "A Brief Narrative of the Journeys of David Thompson," by J. B. Tyrrell,* were commenced in 1799, when he was in his twentieth year. He made carefully estimated traverses of all the routes passed over in his journeys, and also checked them by latitude observations. It may be interesting to follow the record of his journeys in this district. In 1792 he left York Factory and ascended the Nelson river to Sipiwesk lake to spend the winter. On May 28, 1793, he left Sipiwesk House and crossed to Chatham House on Chatham lake, which he places in latitude 55° 23′ 40″, and longitude 97° 44′ 34″ W. This is, no doubt, a point on what is now called Wintering lake. On May 31 he left this place and travelled in a westerly direction to Burntwood river, up which he went to Burntwood lake, and from the western end he crossed to the Missinippi or Churchill river, which he

Reference to David Thompson's work.

---

*Proceedings of the Canadian Institute, 1887-88, 3rd Series, Vol. VI., p. 135.

ascended to Duck or Sisipuk lake, from which he returned and journeyed back to York Factory. In the spring of 1794 he appeared at Buckingham House, on the Saskatchewan river, above Fort Pitt.

<small>Surveys made by him.</small> From there he made a survey of the Saskatchewan river as far as Cumberland House, and thence of a route east to York Factory, by which he followed up and surveyed Goose river and lake, and Athapapuskow lake, then crossing Cranberry portage he followed the Grass river to Reed lake. Here he left one of his associates, a Mr. Ross, probably to build a house, and then proceeded by File lake and the Burntwood river to York Factory. He returned in the autumn to Reed lake to spend the winter at the new house, which he placed in lat. 55° 40' 36" N., long. 102° 7' 37" N. His meteorological register shows that he remained there till May 1797. Shortly after this he transferred his services to the North West Company and moved to a western field of action. In 1804 he again appeared in this district, to build a house at the narrows of Cranberry lake. He wintered at Granville lake, on Churchilll river, and in the spring of 1804 he retraced his steps to Cumberland House. Several minor trips were made to Cranberry lake and one to Reindeer lake in the north, before he went west to cross the mountains.

<small>Compilation of map.</small> The various surveys made by Thompson were compiled by him in 1814 to form a map of the North-west Territories. The original of this is now in the Crown Lands office, Toronto. This formed for many years the basis for much of the geographic detail of our general maps, but it is now being superseded by the more accurate surveys.

Later Explorations.

<small>Later explorations.</small> In 1878 Dr. R. Bell commenced explorations in the valley of the Nelson river and in the next two years the Nelson river, the lower part of the Grass river and parts of the Churchill and Little Churchill rivers. Mr. A. S. Cochrane, in 1880, surveyed the Minago river and part of the Saskatchewan from Moose lake to Cumberland House.

The principal instrumental survey through this district was that of the Saskatchewan and Nelson rivers, made in 1884 by Mr. O. J. Klotz, D.T.S.

The exploration by Mr. J. B. Tyrrell, in the summer of 1896, consisted of traverses of several channels of Nelson river and small streams tributary to it on the west side, the waters of Grass river from Cranberry lake to Paint lake, part of Burntwood river from

Three Point lake to the mouth of Manasan river, Goose river and lake and a traverse of part of the shore of Athapapuskow lake.

In the summer of 1899, the writer made traverses of the upper part of Burntwood river from Three Point lake to its head near Reed lake. Kississing river was also explored by following a route from the north end of Athapapuskow lake to its mouth on Churchill river. The latter stream was also surveyed from above Sisipuk lake, shown on the western edge of the accompanying map, eastward to the end of a long arm running from Nelson lake. In the southern part of the district traverses were made of several lakes to the west of Moose lake as well as the western part of this latter lake, all of which had not been delineated on any of the former published maps.

The series of surveys made by the later parties were in the nature of preliminary traverses, but were done with considerable detail.

General Description.

The general nature of the country shows a rather low relief. The difference in level between the higher part of the upland surrounding Cold lake and the lower portion of the Nelson valley in the vicinity of Sipiwesk lake, is but slightly over 500 feet. *General description of country.*

The most noticeable range of hills is that which crosses the Saskatchewan river at The Pas. This ridge is mainly of glacial origin and is from twenty up to ninety feet high, but situated as it is in a flat country, it forms a very prominent feature. The escarpment formed by the outcrop of the Palæozoic limestones along the southern edge of the valley of the upper part of the Grass river, is another prominent feature. This is in the form of a nearly continuous cliff fifty or sixty feet high facing generally to the north. An eastern face of this escarpment may be seen on Lake Winnipeg, from which it probably continues north.

The area described in these reports and illustrated on the accompanying map is divided naturally into three distinct parts. The largest in area is probably the plateau, which is underlaid by the nearly horizontal limestones of the Palæozoic. Next in importance is the broad valley of the Nelson river and its tributaries. To the west of this is a higher, rough, rocky tract extending west from the outlet of Burntwood lake. In the part underlain by limestone the surface features are very similar to those obtaining in the lake region to the south, but in the depression occupied by the Saskatchewan river and *Division of area.*

the lakes through which it passes, the change brought about by the gradual filling up of the channel by detritus, is a marked feature. In the early history of the river, several lakes were situated along this channel. The ridge crossing the valley at The Pas, at one time held back a large lake, and in this was accumulated a thick deposit of sediment, but as the outlet across the ridge was worn down, the lake disappeared. The river channel across this basin is built apparently above the flood plain. The land on either side is raised but little above the bed of the river channel, and so is subject to periodic inundations. In the country which formed the shore of this lake, it is generally found that limestone beds are not far below the surface, being covered by a light deposit of boulder clay and the lacustrine silt which supports a growth of spruce and poplar.

Old lake basin.

The basin of Moose lake is apparently the remains of a larger one, the southern end of which has been silted up by the river, and through the plain so formed the latter now winds in several crooked channels. The present outlet for the water of Moose and Cedar lakes is by the channel which reaches Lake Winnipeg at the Grand Rapids. There seems, however, to be a possibility of there having been an earlier outlet to the north-east from Moose lake by the channel of Minago river.

East of the outlets of Reed and Burntwood lakes the surface of the country slopes gradually to the east to Nelson river, while beyond that again there is a slight rise to the south-east, forming in this manner a wide though shallow valley or depression running north-east and south-west. This depression is probably continued under the Palæozoic limestones to the south-west. It is quite possible that the limestone beds formerly extended through this shallow depression and joined those bordering the west and south sides of Hudson bay. From the Nelson river westward, the rock, mainly gneisses, are buried beneath a thickness of from ten to one hundred feet of soft gray stratified clay. This clay has rarely been deposited in sufficient thickness to level up the original inequalities of the underlying rocky floor.

Rocks covered by stratified clay.

West of this clay-covered country of the Nelson valley, the underlying rocks emerge at a slightly higher level than in the rest of the district, and form a plateau with rough surface partly barren and unattractive in appearance. The surface deposits are meagre, being limited to a thin sheet of till and occasionally sandy beds in the lake basins.

In the valley of Churchill river a narrow strip of clay is found on which there is a fairly luxuriant growth of small timber. The central and higher parts of the area are nearly barren, but where covered by forest growth it is found to consist of stunted Banksian pine. The general level of this rocky district is over 900 feet above tide, and the greatest elevations above this amount are not over 150 feet. The hills bordering the valley of the Churchill river may, in some cases exceed this, but their greater relative height is mainly due to the great denudation along the line represented by the valley of the above river and again along that of Burntwood lake.

General Geology.

Laurentian.

This term as applied to rocks in Eastern Canada has been given significance as a formational term and the rocks comprising it are thought to form an older series on which the Huronian or earliest sedimentaries rest. Throughout the central part of the continent all the rocks in the Archæan complex, which do not belong to the earliest sediments, are found to have been subjected to such metamorphism, that in the writer's opinion it is impossible to definitely assert at present that any of the various gneisses or schists met with are older than the Huronian, though many bands might be considered to be altered equivalents. In a few instances, between the rocks classed as Huronian and the surrounding granites and gneisses the contact is igneous, and shows granites and gneisses to have been in a state of partial fusion at the time of the folding and crumpling of these rocks.

<small>Laurentian</small>

In the district under discussion it is impossible to map out, under the conditions of a reconnaissance survey these newer gneisses from any that might be supposed to be older. The rocks therefore described and mapped as Laurentian are a series of gneisses composed in part of highly metamorphosed material in close relation with granites and gneisses whose age the writer supposes to be younger than either the series of gneisses noted above or the Huronian.

The original crust of the earth, after its great crumpling and folding subsequent to the deposition of the Huronian sediments, suffered such extensive denudation that the rocks now exposed can be considered as a horizontal section of the crust at a considerable distance below the original surface. The present areas of Huronian are thus the lower parts of such deep folds as penetrated to this level through the harder crust. It may be supposed that the lines on which the greatest movement would take place would be over such areas as

<small>Extensive denudation.</small>

might be still in a semi-plastic condition, and the lower parts of these folds might penetrate areas of uncongealed matter which would digest and remove much of their lower members. The subsequent denudation would reveal at successive depths a lessening amount of the original crust and so it is problematical if any of the original upper beds are to be seen in this area.

In the eastern area, reported on by Mr. Tyrrell, the contacts are more nearly conformable and might indicate that their original relation had not been disturbed in the subsequent alteration to which both had been subject. In the western part of the district there is a marked difference between the gneisses of the area lying north of Athapapuskow lake, reaching to near the Churchill river, and those with which they come in contact in the vicinity of the valley of this stream. In going north the first rock met with after leaving the Huronian area is a granite that gradually becomes foliated and appears as if it might be newer than the Huronian. On reaching the vicinity of the Churchill river an apparently older series is noted, which in some instances is separated from the rocks to the south by wide dykes or areas of an eruptive granite of the nature of pegmatite. Beyond this zone of intrusion there are broad bands of mica schist, garnet-bearing schists and dark gneisses which are a contrast to the generally reddish granitic gneiss to the south.

Huronian.

Huronian.   To the west of Lake Superior the areas which have been referred to the Huronian and of which detailed studies have been made are those on the Lake of the Woods. As the typical section could not be exactly correlated, the group described by Dr. Lawson was called by him the Keewatin, and a lower and more highly altered part, the Couchiching, but it is generally accepted that these constitute in the west rocks representative of the Huronian. The small areas of similar rocks found to the north are thus classed as of the same general series, and evidence is not wanting that many of the beds composing their mass have a clastic origin.

In the eastern part, as will be seen from Mr. Tyrrell's report, clastic rocks, such as quartzites and conglomerates, are associated with basic eruptives and greenstones whose origin is volcanic. Parts of the areas to the west are described in the present report and the same character is found as in the rocks to the east, or in the Nelson valley.

Cambro-Silurian.

The outcrop of these rocks along the western shore of Lake Winnipeg is continued northward and then westward, passing to the south of the chain of lakes on the upper waters of the Grass river. In the southern part of the Lake Winnipeg basin the section gives a thickness of about 270 feet of limestone belonging to the Trenton, but in following the escarpment northward the beds thin out and the lower members disappear. The basal member, a sandstone which rests upon the Archæan, appears on the shore of Reed lake, but it is evidently an equivalent of higher horizon than farther south and is immediately below beds which on Lake Winnipeg are called the Upper Mottled. The section on Reed lake is described by Mr. Tyrrell. The fossils collected by him from the sandstones belong to the middle and upper part of the Trenton. A thickness of less than a hundred feet of Trenton limestone appears above these beds, and a reddish band above these is supposed to indicate a transition to the Niagara.

Cambro-Silurian.

Silurian.

Undisturbed horizontal limestones of about the horizon of the Niagara were seen at several low outcrops on Namew lake to the east of Cumberland lake, as well as on Cormorant, Yawningstone and Moose lakes. On Cormorant lake the sequence observed was as follows. The lowest beds exposed are of a compact reddish dolomite, above which, five or six feet of similar beds weather very rough on the surface. A thin compact dolomite up to ten feet in thickness forms the upper member. These latter beds are shown in better exposures on Moose lake near the old Indian Reserve and on an island to the north. The exposure is in a cliff about thirty feet high showing at the base only two feet of a granular dolomitic limestone and the remainder of thick beds of a lamellar dolomite apparently of coraline formation. The rock is built up in thin plates having a crumpled surface from which many saucer-shaped pieces can be broken out. These are possibly remains of stromatoporoid corals which form the mass of the rock. No fossils were found in these beds but from a few loose fragments of a lighter and more granular rock, pushed up probably by the ice from below, the following forms were observed: Fragments of a Cyathophylloid coral like *Zaphrentis*, *Favosites*, sp., *Strophomena acanthoptera*, *Conchidium decussatum*, *Murchisonia*, two species, *Euomphalus*, sp., and *Gyroceras*, sp. These fossils are all common to the Niagara rocks.

Silurian.

Fossils.

of the Grand Rapids of the Saskatchewan and are from rocks probably in place beneath the section given above.

The Stromatoporoid beds are also exposed along the shore of this lake southward to the outlet at the present Indian Reserve. Slightly higher beds occur near the Saskatchewan river below the "Cut-off" and north of the Moose lake branch, in which small shells like *Isochilina* or *Leperditia* are found. These fossils are very scarce and not well preserved but are sufficient to show that the rocks of Cedar lake which are rich in these forms, continue to the north-west.

Rocks probably Trenton.

On Namew lake the rocks exposed on the north side are probably Trenton but these are overlain by reddish beds and again by white hard dolomites which seem to belong to the base of the Silurian. Fossils obtained at the south end of this lake below Whitewhiley narrows, though many are of new species, show a horizon similar to the Niagara of Cedar lake. The extension of the beds north and east to near the edge of the limestone escarpment is quite probable, since on Cowan river they were followed to near the source of that stream. The eastern edge of the formation is evidently drift-covered, so that the definite outline is hard to trace and it is only in a few localities that it is observed. Westward from Cranberry lake the Trenton probably occupies a narrow band with the Silurian rocks to the south. One exposure on the middle one of the Cranberry lakes shows Trenton beds below a broad red band which is no doubt continued to Namew lake as the transitional beds, and above this again are a few beds of dolomite which are the representatives of the lower members of the Silurian.

Pleistocene.

Pleistocene.

The rocky surface of all this area is scored and polished by the progression across it of a great glacial ice-sheet and in the eastern section evidences are found of a second invasion by another sheet from the north-east. The first came from the north, a part of what is known as the Keewatin glacier. This advanced south beyond the boundaries of Manitoba, and on its retirement or when the accumulation of ice in the north ceased, there was still an active progression in the Labrador ice sheet, and its front ultimately passed the eastern border of the district already scored by the Keewatin glacier. The ice fronts of both the Keewatin and Labrador glaciers are supposed to have met in the region through which the lower courses of the Nelson and Churchill rivers now run, and as the general slope of the land is to the north, the melting of the ice formed a large lake whose western shores were along the face of the escarpment lying to the west of the present Manitoba lake basin, the north-eastern edge being formed by the mass of ice both to the east and north. This lake, of which the present lake basins are small remnants, is described in our reports as Glacial Lake Agassiz.* In the district here mapped, the accumulation of lacustrine material deposited by the waters of this temporary lake, are found in the eastern part to aggregate in some cases as much as one hundred feet in thickness of a fine clay and clayey silt. The discussion of the characters of the deposit will be found in Mr. Tyrrell's report, as well as a combined list of all the observations for the whole district relating to the directions of glacial markings. Many of the observed directions are also indicated on the accompanying map.

Lacustrine deposits.

The western limit of these stratified clays is found to run southward from the outlet of Burntwood lake and enter the basin of Reed lake. To the north of this latter lake instead of a deposit of clay, a sand plain was found, on which were numerous beach ridges formed no doubt at a stage of this temporary lake. Another series of sand and gravel beach ridges were also noted at Cranberry portage. As these ridges are at a much lower elevation than those marking the maximum height of this lake, it must be supposed that the accumulation of lacustrine material was either added to the basin at a late and lower stage, or that during the high and early period this country was still ice covered, and the lake existed only in this locality at a lower level. Over the western portion the rocks are but thinly covered by a glacial till, and on the higher parts, mainly around Cold lake and in the hills near the Churchill river, there is very little covering over the rocky surface Boulders are in evidence, but mainly of gneiss and granite of nearly the same character as the underlying rocks.

Recent.

Evidence of the recent action of the rivers in forming valleys, is not well shown in the western part of the district, as the mantle of clay, or other covering over the harder rocks is there very thin and valleys consequently follow old courses, but in the eastern part many of the valleys of minor streams have formed new channels. Recent deposits in the valleys are of small amount, with the exception of the delta of the Saskatchewan river above Cedar lake. Part of this deposit may have been formed before the recession of the glacial Lake Agassiz, but it is clear that there is an enormous amount of sediment still being brought

Recent changes

---

*Annual Report, Geol. Surv. Can., vol. IV., (N.S.) part E.

down by this stream and the largest part is deposited before the water leaves Cedar lake. From analysis of water from several of the streams in the district a comparison of the amount of sediment contained may be gathered by reference to the following table:—

One Imperial gallon contains suspended matter.

	Grains.
*Nelson River (Sea River falls)	2.565
Reindeer lake	2.02
Churchill river	7.96
Saskatchewan river (near Cumberland lake)	16.60

Peat.

Small deposits of peat are to be found in various places, but the most important, from an economic point of view, is the area north of Lake Winnipeg described by Mr. Tyrrell. Along the valley of the Burntwood river, where it is cut through the thick clay deposit, the general surface of the terrace is quite level. The drainage near the river is general, but back from the edge of the valley, on the more level parts, there is very often a wide expanse of swamp covered by a stunted growth of spruce and carpeted by heavy layers of moss. These swamps may at some future time supply peat for fuel.

ECONOMIC RESOURCES.

Agricultural possibilities.

As the area is situated so far north of the boundary of Manitoba, it might be presumed that much of it is unfitted for settlement, but it is discovered that over a large part there is a good soil, and the evidence of several gardens at various posts show that for all the ordinary vegetables and coarser grains the climate is not too rigorous. Splendid gardens were found as far north as Nelson House, which is in the northern part of the area here mapped. Proper drainage is however needed to bring much of the surface into a condition fit for agriculture. Along the river banks this is evident, for while the strip bordering the streams produces a great variety of grasses, shrubs and trees, a short distance back this is replaced by a swamp covered by moss and stunted spruce. This is more noticeable in the western part of the Nelson valley, where the country is thickly covered by a coating of clay, and the surface is so uniformly level that its gradual slope to the east is not sufficient to drain it. The areas to which it would be possible to introduce a system of drainage, would at first be restricted to a narrow margin along the streams.

*Report of Progress, Geol. Surv. Can., 1879-80, p 79c. and 1889-82, p 6 H.

The north-west corner of the district for present purposes may be classed as without a sufficient soil for agriculture. This may roughly be outlined as being composed of all the country lying to the west of a line from the outlet of Burntwood lake to that of Reed lake, and north of the escarpment which shows the northern limit of the Trenton limestone. In this the surface is rolling and hilly, the rocky ridges having a scanty coating of boulder clay and an occasional thicker deposit in the depressions. It will probably remain the home of the hunter and the trapper. *Area unsuited for agriculture.*

To the south the country underlain by limestone has many of the characters of the northern part of Manitoba. In the valley of the Saskatchewan there are large areas of rich soil formed principally by the river itself which has brought down an enormous amount of silt from the upper part of its valley.

The western part of the valley of the Nelson river is covered by a thick lacustral deposit which reaches west to Burntwood lake and east to the channel of Nelson river. In this area good soil is found in almost every part and where drained would no doubt make fair farming land.

Timber.

In the southern part of the district, spruce of both the white and black species is found of fair diameter, but in going north the size materially decreases. Over the major portion of the rocky country Banksian pine is the principal tree, which though not large enough in general for timber, might in the future be of use for pulp wood. *Timber.*

Peat.

Reference has been already made to the deposit of peat north of Lake Winnipeg, and when a process for preparing this for fuel has been perfected this deposit may perhaps be utilized. *Peat.*

Minerals.

The several large areas of Huronian rocks which are here partly outlined will at some future time be thoroughly prospected, and, as has been the case in nearly all such areas, ores of the useful and precious metals are likely to be found. As it is at present a very hasty visit has shown that many quartz veins and intrusive dyke cut these rocks. and indications of the precious metals are not *Minerals.*

wanting. In the Pipestone area on the Nelson river, mispickel and copper-pyrites are recorded by Mr. Tyrrell, as well as a promising showing of mica on the south side of the Indian Reserve island in Cross lake.

DETAILED DESCRIPTIONS.

The Saskatchewan River.

Saskatchewan river.

From the foot-hills to the edge of the second prairie steppe the Saskatchewan river flows through a country underlain by soft easily disintegrated rocks and therefore it has cut a deep channel. From Nepewin to Thobin rapids the high banks gradually become lower, until below the latter point the river emerges on a low delta plain which reaches to Cedar lake. In its upper part the stream is still actively cutting in its channel, and its waters are at all times heavily charged with the denuded material. In the lower part of the delta the process is reversed and the stream becomes the active agent in filling up what seems to have been a chain of lakes. The uppermost one was probably partly filled while the higher levels of Lake Agassiz still covered this basin. On its recession to an elevation of about 900 feet in this vicinity, it is probable that there still remained a lake whose eastern margin reached to the ridge at The Pas. On the further recession of this former lake, the outlet at The Pas was slowly worn down through the boulder clay and parts of the original lake were drained. The eastern end near the outlet seems not to have been so deeply filled by river detritus. Through the plain thus formed, now winds not only the channel of the main stream, but also several other smaller ones. The course followed by the river of late years is by a channel that has been built up so high above the surrounding plain that at several points other channels have broken out and connected with streams both to the north and south. Latterly, however, one has been opened to the upper part of Cumberland lake, and now most of the water of the river passes through it, and in this way the lake acts as a new settling basin which will rapidly silt up.

Lacustrine deposit not deep.

Near The Pas ridge the depression is but partly filled and shallow lakes appear on either side of the channel. That the lacustrine deposit is not of very great depth is shown in the fact that boulder clay knolls appear as islands in Saskeram lake and also in a ridge crossing the Saskatchewan below Tearing river, on the south side of which an Indian reserve is located. In the low stage of water at the time of Mr. Klotz's survey of the river, boulders appeared at this locality and their occurrence was noted as rare for this part of the river.

The ridge, which forms the eastern boundary of the delta-filled lake above noted, is uneven on its summit and seems to be formed of morainic deposits of varying thickness heaped upon the comparatively even floor of limestone. Exposures of the underlying rocks are wanting where the river cuts through it, but farther to the north on the small island at the north end of the Indian reserve ten feet of horizontal beds are exposed. They are light coloured and, similar to those on Cormorant lake, of the Niagara horizon. On the north side of the river the ridge is higher than any part of it that is visible to the south. The Indian reserve includes the western edge and highest of an irregular hill or ridge running to the north-east. North of this and parallel to it, another ridge runs from north of Watchi lake to the south-western border of Cormorant lake. North of that again another ridge follows the western side of the valley of Cowan river. There is no doubt that south of the Saskatchewan river similar ridges combine to form a strip of high ground to join that which is north of Red Deer river. An outlying hill belonging to the same series was observed on the east side of Cormorant lake.

Glacial deposits.

The lowest gap in the ridge is that at the outlet of the river. Here the stream has worn a channel of considerable depth. On the south side of the plot of ground on which the dwelling belonging to the mission stands, is an old channel, the bottom of which is now at about the level of the ordinary range of high water. The amount of erosion since the river left this channel has been considerable and I am informed that the house mentioned above has been moved three times to prevent its being undermined.

Evidence of erosion.

The following extract from Mr. Otto Klotz's report of 1885 bears on the same subject : " The action of the water in the course of time is well illustrated here. Forty years ago a lad could throw a stone from the banks of the parsonage across the river where it is now fourteen chains wide. Within a few years an island upon which the Hudson's Bay Company's powder magazine was kept, has disappeared. The banks where formerly houses of the company stood (in front of the present post), have been washed away. The same fate is rapidly approaching the parsonage close by." This has since been moved.

On the western bank of this island where the banks are scarped a section of fifteen feet of till is shown. This consists of light-gray un-

Section of till.

stratified clay, containing striated pebbles and boulders. The boulders are of limestone with some of Archæan gneiss and greenstone.

Land suitable for farming.

The high ground here cut through by the river extends only a short distance to the eastward of the Indian reserve and the only land available for farming has been taken up by native settlers. Back from the river-bank there is probably yet plenty of good land. The height of the ridge east from the Big Eddy is estimated by Mr. Tyrrell at seventy feet. The following description from his notes gives particulars as to its surface composition : " In ascending it a terrace is met with at the height of thirty feet and the upper twelve feet is as steep as gravel will stand. The ridge, on the summit at least, consists generally of fine rounded gravel with a few rounded boulders. The material is not well assorted, varying from fine rock-flour to boulders fifteen inches long. The ridge is wooded with Banksian pine and poplar. As viewed from a distance to the westward the summit of the ridge does not appear to be either regular or horizontal, dipping to the north as it does to the south."

Probable beach deposits.

In the interval between this ridge and the one to the north there are traces of the thirty feet terrace as well as several ridges like beach deposits. The rise is very little over thirty feet above the lake level —an abrupt slope at the beach of fifteen feet and then a gradual rise to the beach ridges on the surface. To the east the descent to Atikameg lake is very gradual and the impression is gained that the lake there is at least twenty feet above Watchi lake, but as Atikameg lake is only some twelve feet above Cormorant and Moose lakes, which are at the level of the Saskatchewan river twenty miles above Cedar lake, the difference in level of the two lakes across the ridge cannot be so large.

The hill to the west of Atikameg lake appears to be at least one hundred feet high. Along the eastern face runs what appears to be a terrace of sand and gravel with scarped banks, but as it was viewed from a distance the exact character and height could not definitely be made out. The terrace appeared at about thirty feet above the lake and probably a continuation of that at The Pas. These terraces represent a stage in the level of Lake Agassiz when the waters reached far up the Saskatchewan river and formed a partly inclosed lake. Traces of this terrace or of beaches at a similar level were observed at Cranberry portage and to the north of Reed lake. The beaches on the ridge which separates Cedar lake from Lake Winnipegosis are also at about this level.

Below The Pas the banks of the river again become low and are   Limestone
made up of river deposit, fringed for most of the distance by tall   exposure.
balsam poplar. The first high ground reached by the river is at Pine
bluff, forty-one miles below. In low water, limestone is reported out-
cropping there, and also on the Moose lake branch, a few miles to
the north-east. At Kettle point, twenty-one miles farther down, a
small hill rises ten feet above high water and on its surface boulders
of limestone occur. A small exposure of the limestone is seen in places,
but the surface of the hill is made up mainly of boulder-clay. The
banks below this become noticeably lower as the river is descended
and near Cedar lake are just above the water and fringed with
willow, showing less of the river deposit than in the upper reaches.

Where the banks are built up above the range of ordinary high
water they are of very much the same character throughout. A fringe
of trees skirts the river on both sides and consists principally of balsam poplar (*Populus balsamifera*, Linn.), elm (*Ulmus Americana*, Linn.),
ash (*Fraxinus pubescens*, Lam.) and gray willow (*Salix longifolia*).
The latter grows generally either along the edge of the bank or at a
distance from the river where the land is swampy. In the shade of
the forest, the Saskatchewan berry, (*Viburnum opulus*) or "high bush
cranberry," grows very luxuriantly. The negundo is occasionally met
with, as well as black spruce, though the former is generally confined
to the higher land on each side of the valley on the dryer soil. In
low water the banks rise to over six feet and are apparently level for
a short distance back from the river, but soon sink with a gradual
slope to the level of the marsh or hay flat in the rear. In high
water the river rises to the top of the bank and is then above the
general level of the surrounding country, so that any further rise is
followed by a flooding of the hay lands and an enlargement of the
lakes and marshes adjoining. The water is highly charged with a
very fine sediment which gives it a muddy colour. This is to a great
extent deposited before leaving Cedar lake, and the water issuing to
Lake Winnipeg is therefore fairly clear.

The slightly sloping plain through which the lower part of the river
flows is not so pronounced a lake basin as that above The Pas. A
strip of higher land follows at no great distance to the west of the
main channel. To the east and north the margin of the higher country
is much more irregular. From the north side of the strait at The Pas   Moose lake
the dry ground forms a bay to the north-east, approaching the river   channel.
again a short distance below what is called the Moose Lake channel.
From a few miles east of this a wide low flat runs to the north-east to

the western bay of Moose lake, and running through this is found a small overflow channel draining towards thè lake. Another low plain runs directly east to the south end of the lake, and along the northern edge a line of timbered country is found, in front of which runs another small overflow channel through a chain of lakes emptying just south of the Hudson's Bay Company's post. Between the Moose lake channel and the main river there is probably another small island of high ground, as Mr. Cochrane there records an exposure of limestone. From the eastern bend of the main river a small creek runs to Moose lake. In this the flow is in either direction according to the height of the water in the river and lake respectively. A small stream is reported by Mr. Cochrane as draining from Moose lake to Cedar lake. It is quite possible that this stream flows during high water only, or that the outflow is over a rocky barrier, as even a small stream running through soft deposits would soon cut out a sufficiently deep channel to materially affect the height of Moose lake and thus deflect more of the water of the Saskatchewan river in that direction. Previous to the silting up of the channel through which the river now flows, it is quite probable that the relative elevations of the northern and southern parts of this basin were somewhat different. The northern uplift which is shown in the beaches of Lake Agassiz had not then probably been completed, so that the low country north of Moose lake might have been much lower than at present. The basins of these two lakes were then probably merged in one and for a time drained north-eastward by the valleys of Minago and Metishto rivers. The further uplift at the north to assume its present contour would close off these channels and deflect the river more to the south. Taking this view we can imagine that at first the stream flowed over a wide marsh to the west end of Moose lake, then as the delta grew and the northern uplift was more pronounced the stream went mainly in the direction of the present Moose lake channel. A further uplift caused the breaking away of the smaller branches to the south. The present western channel is a deflection from the higher part of the delta to the margin of the basin, along which less of the river deposit would be found.

Effect of uplifts.

Moose Lake.

Moose lake. This lake is situated to the north of Cedar lake, and acts as an overflow basin to the Saskatchewan river. The direction of flow in the creek leading to it from the Saskatchewan is regulated by the height of the water in each. When the river is low the lake gradually drains out and then is refilled when the Saskatchewan rises. Through a marsh situated to the east of Moose Lake creek there appears to be some drainage also to Cedar lake. The upper part of this, near Moose lake, is blocked by a dense growth of tall reeds so that the channel is lost and the flow distributed over a wide area of marsh.

The present system of water supply is not of a permanent character owing to the shifting of the channel of the Saskatchewan. Older channels formerly flowing to the lake are numerous. The largest of these is one which flows north-eastward to the western arm of Moose lake.

The basin in which Moose lake lies is very flat and the shores rise comparatively little above the water. The contour of the shore line is very irregular and is determined by the remains of portions of a thick bed of flat-lying dolomitic limestone which overlies a porous and easily eroded band forming the floor of the lake. Those portions of the thick bed which were not removed form the main shore. The shallower parts or bays becoming silted up or previously filled by boulder clay have left many stretches with low marshy margins. One of these marshes cuts off Cormorant lake from Moose lake, leaving as a connecting link a sluggish stream flowing to Moose lake. Another low stretch runs north-eastward from the north end and, by report, extends for fifteen miles to the head-waters of Metishto river, a branch of Grass river. The land at the south end of the lake is also level, and except for a few low limestone ridges, is probably all river deposit. Another marshy tract extends from the north-east corner of the lake to the head-waters of the Minago river. *Dolomitic limestone forms shore line of lake.*

The Hudson's Bay Company's post is built on a ridge of flat-lying limestone near the south end of the lake, just to the west of the outlet. The land here is elevated from six to eight feet above the lake and the beds exposed seem to be all of an apparently unfossiliferous limestone, made up principally of thin layers having numerous cup-shaped depressions and dome shaped elevations, suggestive of Stromatoporoid coral formation. A prominent point about six miles north of the post is formed by a ridge of limestone similar to that at the post. On the north side of a large island north of the narrows a cliff of limestone is seen in which thirty feet of beds are exposed. The lower beds show two feet of a granular dolomite capped by thick beds of a lamellar dolomitic limestone which seems to be of organic origin, though no structure is visible to the naked eye The rock is, as before noticed, built up in thin plates having an uneven surface, and many saucer-shaped pieces can be broken out. These are possibly remains of Stromatoporoid corals which form the mass of the rock. The *Hudson's Bay post.*

Fossils.

exposures are very like the cliffs at the Grand Rapids of the Saskatchewan, classed by Mr. Tyrrell as Upper Niagara. The lower members of the formation are exposed near the foot of the rapids and contain as one of the principal fossils the large *Conchidium decussatum*, Whiteaves. No fossils were here found *in situ* in the lower beds, but from a few loose slabs, of a lighter and more granular rock, pushed up probably by the ice, the following forms were observed:— Fragments of a Cyathophylloid coral like *Zaphrentis*, *Favosites* sp., *Strophomena acanthoptera*, *Conchidium decussatum*, *Murchisonia* two species, *Euomphalus* sp., and *Gyroceras* sp. From the top of the cliff, the shore opposite toward the north-west appears low, with a few scattered stunted spruces near the lake, while behind is a low marsh or muskeg, over which can be seen the hills bordering the north shore of Cormorant lake. To the east the shore appears low but covered by spruce and poplar, and is probably underlain by a continuation of the limestone beds here exposed. The north shore is higher and forest-covered.

Timber poor.

The eastern arm running to the north-east from the outlet was surveyed by Mr. A. S. Cochrane in 1880 and 1882. Reference to the map will show its general character and its many islands. Of the north shore, Mr. Cochrane reports that the points are mainly piled high with limestone shingle. Exposures of thin-bedded limestone also occur. Some of these beds are very fine-grained and resemble lithographic stone. The east shore is much lower though also underlain by limestone. Low land extends for a short distance to the south-east and large bays are found behind the islands and points. A high ridge, estimated at 100 feet above the lake, extends to the south-east. Of the appearance of the shores, Mr. Cochrane in his notes says:—'The timber along the eastern shore is, generally speaking, very poor, though occasional large sticks are to be seen from fourteen to twenty inches in diameter, very scattered and far apart. All this shore has been burnt over in patches at different times, which gives the timber a very mixed appearance. The beach is all low and composed almost entirely of limestone shingle, though in one or two places a short cliff of limestone four to six feet high is to be seen standing a few feet back from the water's edge.'

Cormorant Lake.

Cormorant lake.

Ascending the small stream from the north end of Moose lake for about six miles through a swamp, a small crooked bay of Cormorant lake is reached. This bends around the north end of an oblong hill, rising at its highest point to over one hundred feet. It is lying with its longer axis north and south and appears to be partly of morainic origin, similar to that at The Pas. Underneath, limestone in horizontal beds is exposed. This consists of a rough-weathering lumpy do'omite, probably fragmental, overlying somewhat reddish fine-grained Leds. The south-eastern shore of the main body of the lake is low, passing in front of a long strip of swampy land, but as the south end of the lake is reached, where a small stream enters from Atikameg lake, limestone is again exposed in cliffs about ten feet high. The beds are thin and of very hard compact dolomite without fossils, and have a slight dip to the east. To the west, the cliffs rise slightly, and around the bay at intervals are sections of the same rocks with an addition at the base of five feet of hard whitish dolomite, having a very rough surface and showing numerous joints or cracks filled with a more earthy looking but hard matrix. These are very like the rocks on the east side of Namew lake, six miles north of Whitey narrows. At the north-eastern extremity of the lake a stream of dark water flows from a narrow bay lying parallel to the lake to the west, and into this, Cowan river flows from the north, so that this bay, which looks like a part of Cormorant lake, is in reality one of the chain of lakes situated on the above stream. The rocks exposed on the shore to the north of this, in the several cliffs which are there seen, are composed of reddish beds capped by the firm whitish thick beds. The thin beds exposed on the south shore are evidently to be found in the higher country to the north, but do not show in the cliffs near the lake.

Owing to the great similarity of the different beds an estimate of the thickness of the section exposed on these lakes is very difficult to obtain, but the order of occurrence seems to be as follows :—The lowest beds are of a compact reddish dolomite, above which are five or six feet of thick beds weathering very rough on the surface. Above this are ten to fifteen feet of thin-bedded compact dolomite which might possibly be near the horizon of the Moose lake rocks.

Section difficult to obtain.

Atikameg lake lies to the south-west of Cormorant lake. The water is very clear and deep. On the eastern shore the Indians have a fishing reserve, and in the autumn resort there to obtain their winter supply of whitefish. Twenty-four fathoms is reported as the greatest depth for the waters of the lake, and this would indicate that even in summer a good quality of fish would be found if the Indians could set their nets in deep water. Along the western shore is seen the high ridge which also touches the western side of Cormorant lake. This is partly broken through at the south-western end of the lake. Over the lower part of the gap a road has been made to a small lake to the west of the ridge and connected with Reeder lake near the Saskatchewan river.

The appearance of the country to the east is that of a slightly rolling wooded plain, but it is probably partly underlain by horizontal beds of limestone with an occasional swamp. To the south, The Pas ridge is seen to extend eastward a few miles and then die away. The shores are generally strewn with boulders, and near the north end limestone slabs are piled on the beach. At the south-west corner a long point, running parallel to the south shore, cuts off a narrow bay to the south leading to the portage across the ridge. This seems to have been a morainic ridge.

Cowan River.

Cowan river. The ridge which runs west of Cormorant lake continues in a northeast direction and parallel to this along its eastern slope is a depression which becomes shallower towards the north. In this a small stream flows from a swampy tract a few miles south of Reed lake. On its course are several narrow lakes which together with the stream afford a canoe route to the waters of Grass river. The lake at its mouth is but very little over the level of Cormorant lake, but the next above is about eight feet higher than this level and finds its outlet to the lower lake by two streams falling over a steep slope. The fall is passed by a portage through a spruce grove on the east side of the east branch. The lake above is a narrow canal-like body of water about eight miles long, bordered by high banks which, on the west side, show cliffs of hard white and grayish dolomitic limestone similar to the beds at the south end of Cormorant lake. They consist of thick beds which break up into irregular fragments and some of the cliffs are so much broken and shattered that the bedding is not easily made out.

The upward continuation of the river enters through a small gap or break in the shore-line on the east side near the north end. At the

Yawning-stone lake. turn into the river a small projecting cliff has its middle beds so denuded as to bear a fancied resemblence, when viewed in profile, to a face with wide-open mouth—hence the name of the lake, Yawning-stone. The stream is very crooked above this and partly blocked by fallen timber. At the part near the lake some small rapids occur, but fallen timber also blocks it to such an extent that a portage of more than half a mile is made. This leads over a limestone ledge covered by very little soil and a scanty growth of spruce which has been mostly fire-killed. For five miles the river flows through a plain, sloping slightly

Silurian Limestone on Cormorant Lake, Sask.

to the south-west. The banks in the lower part are at first four or five feet high, behind which is another slight rise but this gradually disappears and in a distance of five miles both banks and the rise behind have disappeared and the stream is running in a very shallow channel. The ridge to the west becomes less distinct though the Indians report a continuous ridge to Reed lake. A small lake on the west branch of this stream being situated to the west of the general course of the valley is in a break in the higher ground of the ridge and has on its western shore several exposures of limestone. These beds are similar to those at the base of the Cormorant lake reddish beds, and are evidently near the base of the Niagara, though no fossils could be found to prove the horizon. Above this lake the country is generally flat and swampy and no exposures of the underlying rocks are to be seen, but on the last portage, which is made to the shores of Reed lake, pieces of red sandy limestone are found which evidently come from the shore of the lake and represent the basal beds of the Trenton which must be here very thin. The small hills near the lake seem to be of morainic origin but are probably made up principally of material from the denuded edges of the limestone beds beneath. Trenton rocks are found elsewhere on this lake so that their presence here is certain.

Limestone exposures.

Of the surface features observed on this stream and Cormorant lake it might be of interest to note that the soil of the country near Cormorant lake is thin over the limestone ridges but in the valleys, such as that of Cowan river, a good quality and depth was observed. Between Cowan river and Cormorant lake there is a strip of fair-sized spruce. This is seen to extend up the river as far as the banks are high, but above this the principal tree is tamarack (larch). A long strip of country along the eastern face of the high ridge west of the lake is burnt over and the timber is dead.

Minago River

This stream rises in a low swamp north of the north-east extremity of Moose lake. The Indians occasionally travel by this stream from Moose lake to Cross lake and the route thus followed is practicable for small canoes. A survey of the stream was made by Mr. A. S. Cochrane of this Department in July, 1880, and from his notes the following brief description of the upper or that section above Hill lake is compiled. The lower portion was visited by Mr. Tyrrell in 1896 and his description is given in part of this report.

Minago river above Hill lake.

Route from Moose lake to Minago river.

The route from Moose lake to the Minago river leaves the lake by a small opening in a floating muskeg which fills the northern part of a narrow bay. This stream or opening is barely wide enough to admit of a canoe being hauled through and in a distance of about a mile the shore is reached. A portage of a mile and a half across the height-of-land is reported as very bad and through a knee-deep muskeg. In two places however the portage crosses strips of flat-bedded limestone of about one hundred and three hundred paces wide and about three feet higher than the water in the swamp. The portage ends at a small stream leading directly into a pond of a mile in length. This seems to be the head of the river and the stream draining it is very small and crooked, blocked by beaver dams and overhanging willows so that portages are frequent, including one a mile below the lake of over a mile in length. From Lily lake to Hill lake the river seems to have a wider valley and more water in the channel, though few entering branches are noted. In this stretch there are several rapids and portages which appear to be made over ledges of flat-lying limestone with a slight dip to the south. The highest exposure on this stream appears to be at about four miles below Lily lake and is of an unfossiliferous limestone. This is probably a bed situated near the top of the Trenton or the base of the Niagara formation.

Burntwood River.

Burntwood river.

The upper part of the Burntwood river lies in a rocky depression on the surface of an uneven plain which is situated between the valley of the Churchill river and the wide and rather shallow depression in which the Nelson river runs. In that part which lies above the point at which the portage-route from Nelson lake joins the river, upward, the fall in the river is trifling and occurs mainly on the short stream near Loonhead lake. Below the point mentioned, the stream enters the country sloping toward the Nelson river. From Reed lake to these waters there are three routes. From the eastern end of Reed lake a small stream may be ascended to near File lake and a portage made to it, or to another stream entering at the northern side of Reed lake which flows from the west side of Methy lake and affords a fair route with a short portage. The third route is by a direct portage from Reed lake to Methy lake. This starts from a sandy bay on the north side of Reed lake and runs directly north through a sandy country covered in the highest parts by Banksian pine. The trail ascends gradually to gain a series of beach ridges which run parallel to the east shore of Methy

Banksian pine.

lake, but from its south end they spread out towards the east to become parallel to the north shore of Reed lake. At Methy lake it is found that this series of beaches, which are along the margin of a plateau of sand lying to the east, rises in steps to a height of fifty or sixty feet above the lake. By a short excursion to the east it was found that for two miles the surface was nearly level but had a slight fall to the east from the highest beach which was only a quarter of a mile from the lake.

These beaches appear to have been formed when the Labrador glacier made a dam across the valley of Grass river and a lake occupied the basins of both Reed and Wekusko lakes. An outlet seems to have existed along the valley of Methy lake. Not enough is yet known of the general disposition of the lacustral deposits and of these beaches to warrant any definite statement, but they may possibly be traces of glacial Lake Agassiz. The basin of Methy lake is along the strike of the Huronian schists. At the south end, as before noted, the valley reaches Reed lake and at the north end joins the depression filled by File lake. The outlet is near the north end where a small stream trickles over dark-greenish schists containing many needle-like crystals of hornblende apparently of secondary origin. A short portage of 150 yards long is made to File lake. Along the western shore the rocks strike north-and-south and are mainly hornblende-schists which at the north end become very much contorted and the strike is there bent through an angle of 130° so as to run south-west and north-east.

Beaches formed by Labrador glacier.

These beds are exposed again on the stream above Loonhead lake and are there found interbedded with light-coloured gneisses or granites. Instead of the evenly fine-grained beds of Reed lake which are smoothly glaciated, the surface of the rocks of File lake are roughly weathered, owing no doubt to the partial decomposition or recrystallization of the hornblende constituents. On Loonhead lake the schists give place to a wide belt of granite included in the Laurentian and after an interval of muskeg, through which the stream passes, the gneisses of the Burntwood lake region make their appearance, striking north-north-west and south-south-east with a few local deflections as far down the river as the beginning of Burntwood lake. Leaving Loonhead lake the stream runs northward and at a mile from the lake falls ten feet over a ledge of gneiss, past which there is a portage of 200 yards on the south side. The rocks are fine-grained, thin-bedded dark gneisses. Below this the river makes a long bend round by the east and approaches the same gneissic ridge, over which it crosses

Belt of granite.

a second time, falling, in two rapids, about six feet. From here on to Burntwood lake the navigation is hardly interrupted, the channel narrowing occasionally so that there is a noticeable current, but it seems to consist of a succession of narrow lakes bordered by rocky banks, and as Burntwood lake is approached these rise in to hills. Very little Banksian pine was seen. A few groves of small spruce and poplar occupy the low parts where there is a little clay and sand between the rocky knolls.

The surface of the rock is everywhere glaciated, showing striæ running S. 20° W. At a lake about eight miles below Loonhead lake where the river makes a jog to the east for three miles, the central island and a long point reaching out from the south-east, are both found to be composed of light fine-grained dolomitic limestone, dipping along the eastern edge, towards the north-east. As the beds are not all standing in this position but are more nearly horizontal at the south-west side, it is possible that beneath these are sandstones of a friable nature which have been denuded so that the beds have fallen down. There appeared to be no fossils in the beds with the exception of a few broken crinoid stems, so the exact age could not be decided, but in their general appearance they resemble the beds exposed on Cumberland lake which are of Niagara age. The limestone is fine-grained but pitted by numerous small cavities, possibly impressions of salt crystals. This outlier of limestone is the only one known in this district at any great distance from the general outcrop of the Silurian and Cambro-Silurian rocks. The lake in which they are found is generally called Limestone Point lake.

Age of limestone beds doubtful.

From here onward for about eight miles the river runs north-northwest, following the strike of the rocks, which become garnetiferous and generally of a dark colour at the end of that distance. It then turns N. 20° E. till it enters the main body of Burntwood lake, flowing through a succession of narrow lakes connected by deep channels. In this distance the channel cuts occasionally across the strike of the rocks. Those in the upper half of the distance are running N.W. and S.E., becoming more and more contorted, until at the middle of the course, light reddish gneiss and granite appear with included fragments of the darker rocks. The reddish rocks seem to have been in a plastic state at a later period than the dark gneisses.

The strike of these later rocks is about east-and-west and they continue north to the main body of the lake, approaching which they are seen to be broken into by large dykes of flesh-red granite.

Burntwood lake is unlike many of the other lakes of the district as it is but a narrow channel, or rather three channels meeting to form a Y. The southern branch may be said to run as far as within two miles of Limestone Point lake, as the first current is there met. The western arm reaches to near the waters of the Churchill to which there is an old portage road. This part is more regular in shape than the southern one as well as wider, probably because its course lies nearly along the great break indicated by the flesh-coloured granite dykes also noted on Cold lake and part of Churchill river. Near its western end it breaks across through some of the ridges and continues on in the same direction but on a course three miles to the south-west. In this latter part the lake is bordered by high rocky hills The eastern arm broadens out and several large islands are found. The rocks of this eastern portion and also down the river (as the outlet is from this branch) as far as the first rapid, are all striking nearly north-and-south or about N.N.W. and S.S.E. They are gray and dark garnetiferous gneisses which show in high ridges on both sides of the valley. The channel runs about north-east but is deflected back and forth along the direction of the gneiss ridges. The outlet of the lake may be said to be situated at the first narrows where there is a strong current. This point is only about four miles below the widest part of the lake and about twelve from the inlet of the south branch. From there to the portage from Nelson lake, the river is very much of the same character as that of the south branch-narrow and with a deep channel flanked by high ridges of gneiss.

<small>Burntwood lake.</small>

<small>Rocks of eastern portion.</small>

From the above mentioned portage the river turns to the east, and begins its descent to the basin of Three Point lake. This basin is situated, by an estimate of the fall in the river, about 150 feet below the level of Burntwood lake. In this part of its course the character of the surrounding country is of a totally different character owing to a deposit of clay of lacustral origin which is spread over the eastern side of the slope lying to the west of the Nelson river.

<small>Clay deposit of lacustral origin.</small>

Near Three Point lake the deposit is of considerable thickness as the river has cut out a deep valley which decreases in depth up to Burntwood lake. On the Burntwood lake basin there is but a thin coating of soil of any kind. Small terraces between the rocky ridges appear here and there but as the river is descended these are more pronounced and, as noted at the portage from Nelson lake, form a definite terrace at five or six feet above the river. The slope of the underlying rocks is apparently a trifle steeper than the surface of the clay as the high ridges of gneiss, which form a prominent feature of the western part of the district, are here partially buried by the clay and the summits only appear at a distance from the river. At the various rapids the underlying rocks are generally seen, but elsewhere rock exposures are infrequent. The first portage is at a fall of eight feet. The trail road is on the south side and is called Carrot portage. It is through a fairly heavy bluff of poplar, small spruce and Banksian pine to a small lake or pond at the foot of the rapid. Shortly below this, the stream enters a rocky gorge, through which there is another fall of eight feet. There the principal tree is the Banksian pine and the hills on either side seem to be fairly well covered by it. The rocks at the fall are a reddish gneiss striking north-east and dipping 20° to the north-west. Below this fall there appears to be a belt of land with good soil skirting the river for some distance. Occasionally a rocky point protrudes from beneath the clay, though as a rule the banks are fringed with willow indicating alluvial soil.

Timber all small.

The timber near the river is mostly poplar but a short distance back it is Banksian pine and spruce, but all very small. Flathill portage, the next below, is at a fall of ten feet. The granite ledge which crosses the river here is seen on each side rising in a high ridge fifty feet above the clay terrace. For a short distance below Moose portage the valley is not deep, but at Clay portage the stream falls twenty-five feet into a much deeper channel which for six miles has scarped banks. The channel then widens out and the stream emerges on what appears to be a lower terrace. Below the fall at Clay portage the rock is a reddish gneiss with bands of mica-schist and garnetiferous gneiss lying nearly horizontal but with a slight dip to the north-east. The banks there are about forty feet high and are composed of sand and gravel with a bed of clay on the surface. For a considerable distance below this the river flows through a fairly level country with here and there a boss of the harder rocks protruding through the clay plain. The mantle of clay here covers all the interval between the greater ridges and the river which in flowing down the slope to the east runs more or less across the direction of these ridges, so that when the valley is worn down to any extent, rapids are nearly always found situated in line with these rocky hills. The stream is more or less a succession of still stretches with deep quiet flow, and shallows and rapids, generally at the points as above noted. Many of these ridges form isolated knolls with their longer axes running in the direction of the strike of the rocks. One of these is noted just above the mouth of Muddywater

river and on the line of its axis rapids appear in the river. At Drift- Falls at Driftwood rapid.
wood rapid there are two falls of four and five feet respectively over
red granitic gneiss, striking N. 20° E. and S. 20° W. A mile below
this, at Grindstone portage, the river again falls over beds of similar
red gneiss. There is very little fall for the next four miles, or until it
passes along the west side of another rocky ridge. Then it turns to
the east and there are four falls at intervals of less than a mile, making
a descent of about forty feet. The first is a fall of seven feet, and the
second of eight feet; the third, Leaf rapid, is a fall of eight feet, and
the last, Gate rapid, of seventeen feet. At the first of this series the
rocks are reddish granitic gneisses with a few bands of included fragments of darker gneiss striking north and south. At the second, the
rock is a contorted garnet-gneiss, followed on the east by a porphyritic
granite-gneiss. At the third, the rock is similar to the second, and
the same rocks continue to the fourth. The river below Gate rapid
enters a deeper valley and makes a bend to the north. The banks are
sand and clay, and before Three-point lake is reached, they have risen to
about thirty feet. In this interval several rapids are situated but the
portages are all short. The last rapid to be passed before reaching the
lake is called Moose-nose rapid, where the channel is constricted by an
out-crop of gneiss which forms on the east side a boss of rock bearing a
rude resemblance to the nose of a moose—hence its name. Below this
the channel broadens out and the current is sluggish, except at a few
points. Near the lake the valley turns to the north-east and joins the
basin in which lies Three-point lake. Banksian pine is growing thickly Timber.
on the edge of the valley, but in places large groves of spruce and
tamarack appear in the lower parts and along the edge of the stream
are groves of black poplar and birch.

Athapapuskow Lake.

From the north shore, which is profusely dotted with islands, a long Athapapus-
bay runs to the north. The shores and islands in the north-eastern kow lake.
portion of the lake consist of green Huronian schists and fine-grained
massive gabbro. About five miles south-west of the head of the river,
this greenstone is overlain by Trenton limestone which soon forms a
low escarpment a short distance back from the beach. The southern
end and part of the north-western shore were not visited. On the
south-west shore considerable areas are covered with large white spruce.
The route to the headwaters of Kississing river is by a stream flowing
into the north end of this lake. To reach it the north-east shore was
followed from the outlet. The main body of the lake stretches to the
south-west and is generally free of islands.

After passing a prominent point a mile from the outlet the first rocks noted on the north shore are of a dark-green squeezed eruptive; the lines of stratification or foliation, though indistinct, run north-east and south-west. The shore is fringed with small spruces and occasional birches. At five miles west of the outlet, on passing through a narrow channel behind an island two miles long, another bay is crossed which runs to the north-east with the strike of the rocks, and to the west on one of the islands, is seen a cliff of limestone capping the central part, the lower margin of the shores being of Huronian greenstone. Westward from there the north shore is said to be capped by similar limestone. Passing behind another island by a narrow strait, a much larger opening is entered, but at the entrance, two small islands are observed composed of a light reddish-coloured rock. This is found to be a granular granite, partially stained by greenish-coloured minerals, probably from the nearby contact with what seems to be an intruded Huronian mass. The rocks along the shore of the larger islands just passed are more crystalline than those first seen and appear to be massive. The colour is a dark gray-green, weathering brownish.

Huronian rocks.

The first of a group of islands half-way across the bay to the north is made up of a dark-green squeezed and altered granitic gneiss with the foliation bearing N. 38° E. The rocks of the islands and on the point north of this are of a soft fine-grained greenstone, containing many rusty specks and small masses of calcite.

On the point to the east of the entrance to the next bay are green schists striking N. 29° E., but most of the rock in the vicinity is massive in structure. The hills around this bay are partly bare and seem to be of rounded bosses of greenstone. The only timber seen is spruce, with a few birch trees.

Indian camp.

In the strait, a small level patch on which there is some soil, is the site of an Indian camp where there are a few graves carefully preserved and neatly enclosed by a wooden fence. This camp is occupied each year by a few families whose hunting ground is at the height-of-land to the north.

The rock is a pseudo-conglomerate formed most probably by pressure and shearing. The matrix is a fine-grained green schist inclosing angular and sometimes ovoid fragments of a coarser crystalline rock lighter in colour. In a few cases the latter consisted of fine-grained greenstone apparently, broken up dyke material. The foliation was N. 20° E.

The Pine-root river, which was ascended, empties into the west side of this bay two miles from the entrance. The mouth is hidden in a grassy flat and the valley through which it flows is not a prominent feature, as it is crooked and narrow. It drains three closely connected lakes at elevations estimated at 60, 65 and 75 feet respectively above Athapapuskow lake. The lower one is only about four miles from the mouth of the stream. Most of the fall occurs near the outlet from the lake where several cascades make a descent of forty feet. Lower down smaller rapids are met, but these are each not over five feet in height.

The rocks noted on the river are mainly at the several portages. Near the mouth the stream flows along the eastern face of a ridge of greenstone running with the strike, nearly due north and south and at several places on the faces of some of the more abrupt parts the rock is seen to be glaciated, the striae running down the valley. The rock showing at the foot of the lowest rapid is a black or dark green quartz-porphyry. The particles of quartz are small and the matrix very fine grained. At the upper end of the rapid the rock is a dark quartzite-conglomerate with a few small pebbles of a bright red jasper. This band lies to the west of the quartz-porphyry and the river crosses it again a short distance up. Irregular veins of a milk-white quartz appear on a boss of rock on the west side of the fall but they seem to be segregations and not fissure veins. At this portage a terrace of sand and gravel is crossed which is about fifteen feet above the water. *Rocks of Pine-root river.*

The strike of the rocks in this part of the valley is very nearly north- *Ridge of conglomerate.* and-south, and the first two rapids cross and recross a band of con- glomerate which to the north and south forms a distinct ridge. The stream cutting through this from the eastward leaves a small basin in which is a narrow lake. From the north-east corner of this lake to the larger one above, the rocks are all green schists striking along the course of the stream or about N. 20° E. and for most of the way the stream runs between high ridges of the schists. At the outlet from the lake the valley terminates and the water descends about forty feet in a series of cascades. A portage of $\frac{1}{4}$ mile on the east side passes over a ridge of greenstone and green schist striking N. 12° W.

The lake is not above two miles in length and scattered through it *Wabishkok lakes.* are several small islands lying in rows parallel to the strike of the schists. At the north-east corner a small round lake is separated from the main body by a ridge of dark-green rock, partly schistose, over which the water of the upward continuation of the stream flows, making a fall of two or three feet. At the east side across the pond is the mouth of a small quiet stream which connects with the middle lake.

<small>House occupied by Indians.</small>

There is a small well-built house on its banks, the winter home of one of the families of Indians who hunt in this vicinity. The middle lake is not as large as the lowest but longer and rather narrow. About the middle of the distance up this lake to the inlet of the stream above, the strike of the rocks, which for a short distance had been difficult to make out owing to their massive character, was clearly observed to be nearly at right angles to that on the lower lake. They are here striking east-and-west, nearly vertical, but at the inlet of the small stream from the upper lake, where there is a short portage, there is a semblance to a dip of W. 38° S. <40°. The rock is dark-coloured, massive in structure and very much weathered or pitted on the surface. Some of it which is dark gray is soft enough for pipestone.

The stream from the upper lake is only some 500 yards long and as it leaves the lake it falls ten feet over a ridge of dark serecite schist striking across the stream in a direction W. 28° S. or E. 28° N. dipping at an angle of 70° to 80° southwards. On the upper lake the Huronian greenstones come in contact with Laurentian granite-gneiss and the line of contact occurs along the longest diameter of the lake.

<small>Islands of Huronian rock.</small>

All the large islands appear to be of Huronian rock. On one near the contact this appears to be partly recrystallized giving the rock the aspect of a diorite. The surface is rough and many grains of quartz appear throughout the rock.

On the north-west side of the lake the rock is a fine-grained granite with few inclusions or patches of dark hornblende-rock in it. There is a slight foliation running N.E. and S.W. A crooked channel leaves the lake at the north end and runs with a general north-west direction for two miles, ending in a small round pond into which only a rivulet runs. This is just to the south of the height-of-land.

Kississing River.

<small>Kississing river.</small>

The portage over the height-of-land is three-quarters of a mile in length, and the direction followed from the south side is generally about N.N.W. At the south end the rocks are reddish-gray gneiss striking north-east. The northern half of the portage is through muskeg, but rocky ridges occur here and there showing gneisses and schists dipping steeply to the north-west. Before reaching the north end the trail descends a steep hill, the boundary of a basin in which lies a small lake. Muskeg extends from this hill out to the edge of the water and the outer edge is more or less a floating bog. It was estimated that this lake is lower than the one south of the height-of-land, probably ten feet. Another portage is made from near the eastern end, where a small stream trickles out to the north, to a pond lying ten feet lower. This is connected by a crooked channel through a grassy and muskeg flat with the east end of an arm from Kisseynew lake, the first of consequence on Kississing river.

The timber here is all very small and chiefly Banksian pine on the ridges with stunted spruce and tamarac in the muskegs.

Kisseynew lake, of which only a part was surveyed, seems to occupy a long basin or hollow lying along the strike of the gneisses which outcrop on its shores and islands. The islands are mainly the summits of long ridges of gneiss which at the north-east end also form long finger-like bays. The rocks at the south of the first bay are light-reddish gneiss, striking east-and-west and dipping north at an angle of 70°—80°. On the lake, however, they seem to run nearly north--north-east and south-south-west. On the island near the north shore the rock is a grayish green massive granite with a slight foliation north-east and south-west, while along the north shore of the same island these are overlain by dark-gray hornblende-gneiss and schists which strike N. 70° E., dipping north-east 45°—60°.

Kisseynew lake.

The river leaving this lake is wide, and with sluggish current. For two miles it passes through a muskeg through which here and there appear ridges of rock. It was quite evident on entering the stream that the small creek by which we reached the lake did not form a very important branch, but that the head-waters must be situated much farther to the west and a stream of larger size will probably be found to enter at the western end of the lake.

At the outlet from Kisseynew lake there is a small fall of three feet, above which is a gray gneiss striking east and west and nearly vertical. The river is wide and deep for nearly two miles down, but then suddenly turns to the north and runs through a small break in a ridge of gneiss, falling eight feet to follow for a short distance its previous easterly course. At this fall the gneiss is striking E. 10° S. and in it are granitic inclusions or segregations of feldspar and quartz drawn out into long stringers. A little soil here covers the rock, and below the fall the river banks are found to have a low terrace of five to eight feet of sand. On the ridges Banksian pine is the principal timber and in the valley below to near Cold or Kississing lake this tree is found in tall grooves on the sandy terrace. A few scattered spruces are also seen.

Kississing Lake.

Kississing lake.

Nearing the lake the valley broadens out to half a mile between the line of trees on either side and the stream winds in a very crooked course through grass and reeds. At the south-west end of Kississing lake the stream falls into a long bay, the shores of which are low and the water a dark yellow. No timber, except small spruce and Banksian pine, is seen on its shores. On the first island the rock is mostly light-coloured pegmatite, containing fragments of schist. Light-coloured rocks appear on the east shore, probably belonging to the same intrusion. North of the island gray-gneiss appears in nearly horizontal, wavy beds, which have a slight dip to the north-west.

In the strait leading to the larger part of the lake the gneisses containing a few veins of pegmatite, dip about 10° north-east and on the bank there is exposed about six feet of sand with a light soil on the surface containing a few boulders. Entering the larger part of Kississing lake it is found to be so thickly dotted with islands that any extended view is limited to an occasional glimpse between them. The hills on the east side are very thinly coated with timber and a few of the islands have a little spruce on them.

Lake traversed to outlet.

The lake was traversed on a direct line through the islands to the outlet. As the country to the north-west is rather low the mainland in that direction could not very well be made out. On the east a range of hills, bare and rocky, forms the east shore and the limit of the lake in that direction was seen to be about four miles, keeping nearly parallel to our course. As the several rock-exposures noted are all on the islands and not easily identified a few notes are given on these localities with their distance from the outlet.

At 10·8 miles from the outlet, the rocks are gray micaceous hornblende-gneisses dipping to the north-north-east at angles of 10° to 20°. A third of a mile north-east on a large island a wide dyke of pegmatite breaks through hornblende-gneiss and schist, which in places are liberally charged with pyrites. The contact with the intrusive mass has oxidized some of the pyrites so that the surface in the vicinity is stained with rust. The dip is not constant and the beds are somewhat wavy, but the average inclination is W. 30° N. at an angle of 40°.

Three miles from the outlet the rocks are light-coloured gneisses containing quartz and very little feldspar with specks of biotite. Small garnet crystals also appear in a few of the beds which are dipping N. 30° E. at an angle of 30°. At the outlet, the gneisses are gray in colour and dip N. 20° E. at an angle of 20°. Kississing river below this lake is much larger than the stream above. Its course is at first in a northerly direction, passing over several ridges of gneiss with sandy terraces near the stream. The course of the river is then deflected more to the east, and at half a mile falls eight feet over a ledge of gneiss. Below this for two miles the stream flows due east between ridges of gneiss parallel to the strike, falling at last over several small rapids. The central one of these has a fall of over five feet and a portage is made for seventy-five paces past it. A sudden turn to the north reveals another fall of ten feet over a rocky bed, past which it is necessary to portage for a distance of 400 yards. Clay is observed on the portage road, which rises ten feet above the water at the upper end. The upper surface has the appearance of a terrace but mostly of sand. *Kississing river below lake.*

This is another small rapid with two feet fall a short distance below this, when the valley is seen to open out and the view ahead is of almost bare rocky hills, a continuation of the ridge forming the eastern boundary of Cold lake. As the river approaches this ridge it is deflected to the north-east and soon, bare and rocky hills appear, on the north side the timber having been burnt over. The course of the river from the edge of the ridge to Takkipy lake is in the form of a long curve to the north-east. In this distance one rapid of four feet was run at five miles from the lake, where the course of the river crossed the strike of banded red and white gneiss and a few thin belts of mica-schist dipping north at an angle of 10°. Below the rapid the course of the stream is with the strike of the rocks. The hills on either hand seem to reach an elevation of somewhat less than one hundred feet. The whole country appears to be covered by small Banksian pines of four or five years growth. *Country covered by first of small Banksian pine.*

A ridge of red granite rises on the north side of the valley three miles from the lake and is probably the same as that which crosses again below the lake. The valley broadens out and the stream flows with a very crooked course and little current through grassy flats before reaching the lake. Between the ridges on the side of the valley small terraces of sand and gravel with a little clay, rise to fifteen feet above the water and the Banksian pines become much taller in patches and are associated with a sprinkling of spruce.

Takkipy lake lies in a basin surrounded by nearly bare hills. The outlet flows from the north end of a narrow arm where the river breaks through a ridge of reddish granite-gneiss, with a fall of eight feet. A portage of one hundred paces is made over heavy bedded gneiss dipping north-east at an angle of 30° to 40°. The valley below this fall continues in the same northerly direction but soon narrows to a canyon with steep rocky sides through which the river falls fifteen feet in a distance of two hundred yards. At the lower end of the portage is a small thicket of poplar and spruce, but the timber on the higher parts, both at the portage and on the surrounding hills, consists of small Banksian pine only.

From below the fall the northern edge of the high rocky country which surrounds Takkipy lake runs to the west, but on the east it is not so definite, as rocky ridges extend to the north. From a valley on the north a small branch enters and half a mile below this the river falls about four feet in small rapids. The rocks are quite massive and appear to be nearly horizontal but dip slightly to the north. After passing a small round hill on the west side, a branch, the largest yet seen, joins the stream from the west. From this to Beaver fall the stream is fringed with rushes and the current is sluggish.

Island of garnetiferous gneiss at Beaver fall.

At Beaver fall the river divides and falls fifteen feet nearly perpendicularly on each side of an island of garnetiferous gneiss. A portage is made across the island over bare rocks to the foot of the fall, a distance of about twenty-five yards. The rocks are nearly horizontal. Before reaching the portage they appear to be dipping slightly to the south but at the foot they are dipping slightly to the north. The beds are of garnetiferous gneiss interstratified with a light red granite-gneiss. Protected surfaces show glacial striae running S. 39° W. On the banks there is some good soil on which is growing fair sized timber, mostly poplar. The area of good land in this part of the valley must be small as the rocky hills are but a short distance back from the river.

The lake into which Kississing or Cold river enters is on the same level as the Churchill river, with which it is connected by a narrows at Shaving point. It occupies a deep rocky valley dotted with many islands. Along the sides of the valley and covering the summits of the islands is found a deposit of clay in which are noticed many small concretions somewhat similar to those from the clay of the Nelson River valley. It is confined here, however, to a narrow strip along the valley of the Churchill.

The character of the rocks in a great measure contrasts with those on the Kississing river. The evenly bedded gneisses are here broken into by large dykes of a salmon-coloured granite and the dip is increased to a high angle, becoming almost perpendicular. At the south end of the lake the salmon-coloured granites are much in evidence and form large patches on the prominent points. Near the mouth of Kississing river they are seen in the cliffs to be generally interstratified with darker gneiss. A short distance to the north there are many examples of beaded gneiss in which the darker rocks are very much altered and drawn out in irregular forms. Many of the beds are very much seamed and broken by dykes of the pegmatite and the fragments show greater alteration and squeezing. The direction of the dykes is about parallel to the strike of the gneiss or W. 30° N. and E. 30° S.

Character of rocks.

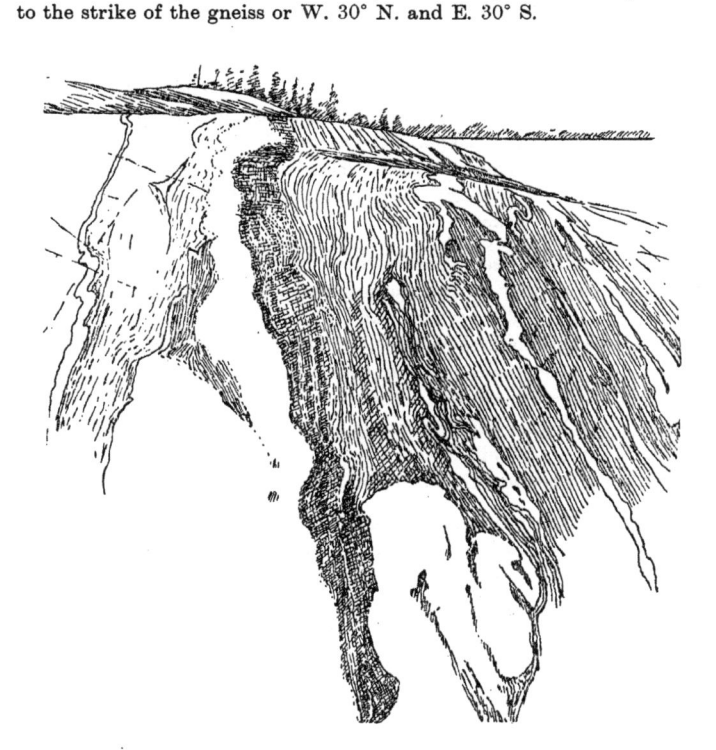

*Pegmatite dykes and squeezed Gneiss and Schist Churchill River.

Churchill River.

Churchill river.

The uneven nature of the rocky floor of the valley is seen in the many island-studded lakes along its course. From Shaving point westward for five miles the channel is generally wide with a moderate, even current, but where there are contractions the current becomes stronger and at two or three places for short distances reaches four miles an hour. On the lake the gneisses seem to be running east-and-west, and are generally studded with garnets. The glaciation here is all to the south-west.

At the foot of Pukkatawagan fall the rocks are contorted garnetiferous gray gneiss which seems to have been so much crumpled and contorted as to have lost all general strike. On the lower part of Kississing river the included fragments in the granite dykes were less altered than here and preserved a rude alignment. Here, there has been more movement in the magma and greater alteration. At the fall there is a beautiful cascade of twelve feet broken by an island in the centre. On the north side two other falls occur on another channel which runs to the north of a large island. The portage is on the south side, 340 paces long, mostly over bare rock. This is a dark-gray gneiss dipping north at an angle of thirty degrees, broken into in many places by large red granite dykes.

Portage over high terrace.

Continuing in the same general direction for another mile the stream falls through a rocky cañon in a long rapid, with a total fall of about twenty-five feet. The portage road on the north-east side runs over a high terrace and point of rock for 600 paces coming down to a bend in the river below the rapid. On the terrace the soil is clay with boulders and upon this and the slopes near the river are small groves of poplar. The higher parts are sandy and thinly covered by Banksian pine. Many of the hills above this level are quite bare. The rocks at the fall are horizontal thick-bedded red granite-gneiss.

Near the Churchill river the valley is almost free of timber, except a little on the slopes of the hills and near the mouth. The rocks are massive granite-gneisses with a slight dip to the north. At the mouth the rocks are massive granites with contorted inclusions of darker gneiss. Wherever there is any foliation it is east-and-west with a high dip.

Trading post.

Above the fall the river expands again into another lake which continues on to the west for six miles and then turns to the north-west for about the same distance. On the north side, near the bend, the

Hudson's Bay Co. have a winter trading post. There are several houses and a Roman Catholic mission building. These are all built on the surface of a terrace of clay ten feet above the lake, and although there did not seem to be any gardens attached, there were several potato patches on the islands in the vicinity. The rocks above the fall are light-gray massive gneiss, and on the islands two miles to the west they are mostly of the light salmon-coloured pegmatite. At the post the rocks are a light-gray gneiss, nearly horizontal, but with a slight variable dip to the north. On the part of the lake running to the north-west the strike of the gneisses follows the direction of the lake. About the centre of this part of the lake the strike is nearly north-and-south, with a dip of only 30° to the east. At the western end of the lake the strike has again changed to east-and-west. *Roman Catholic mission.*

The depression filled by the lake, thus seems to follow very closely the line of the strike of the foliation of the gneisses as well as that of the great break or breaks now filled by the light-reddish granite.

On Bonald lake the rocks are mostly of the light granite with inclusions or streaks and patches of dark gneiss, running in many directions. The hills are clothed mostly with Banksian pine, but occasional groves of black spruce with a few tall trees of white spruce are seen. At Bloodstone fall gray gneisses running east-and-west and dipping to the north are cut in the vicinity of the portage trail on the south side of the river, by wide dykes of a coarse red granite. The name of the fall is possibly given on account of the red granite. A few garnets are to be found in the gneisses but these are not so prominent or large as at Pukkatawagan fall. *Banksian pine.*

Sisipuk lake occupies the upward continuation of the valley in which the river flows from Bloodstone fall to the inlet to Pukkatawagan lake. On the south, skirting the shores of the lake, rises a prominent line of hills. To the north between the lake and the river the country is not so elevated. The rocks at the east end run east-and-west with a slight dip to the north. They are mostly of gray gneiss with lighter-coloured streaks of granite. Towards the middle of the lake the rocks are garnetiferous gneisses. On the islands leading northward to the mouth of the river the rocks strike north-west and south-east, dipping to the north-east, and show many examples of the striped rocks such as are seen in the lower part of the river. A short distance north of the lake the river divides passing around a large island. On the smaller branch the fall is at two and a half miles from the turn. This is a chute about forty feet wide. Past this the port- *Sisipuk lake*

13—FF—4

age trail is over a clay terrace rising ten feet above the water and lying between two hummocks of rock. Above this fall the channel broadens out, and running to the west, joins the main body at the south end of Loon lake.

The gneisses here trend nearly north-and-south, with a high dip to the east and many granite veins cut through them. On one of the small islands in the channel the exposure has the appearance of a patchwork of dark fragments inclosed in a light granite.

Above Loon lake the river again passes on each side of a large, high island for eleven miles. The channel on the east side is narrow and in a few of the narrower places a slight current is observed, otherwise all this part is of the nature of a straggling lake. The strike of the rocks gradually swings around from a north-and-south direction on Loon lake to east-and-west in the narrow channel above mentioned, and on Mountain lake, from which the portage to Sisipuk lake starts, the strike is south-west and north-east.

Portage to Sisipuk lake from Mountain lake.

The portage to Sisipuk lake measure three-quarters of a mile through gap in the hills, leaving Mountain lake at a terrace of sand fifteen feet high, on the surface of which there is some good soil. Most of the distance is through small spruce to a marshy inlet from the west end of the lake. Above this the river issues from a narrow gorge in the range of hills which runs along the south side of Sisipuk lake. The usual course taken in ascending this part of the stream is to follow a channel on the east parallel to the main river, and portage over a ridge to Doctor lake above. Descending from Doctor lake, the rapids may be run by keeping close to the shore on the west side. At the foot of the rapids the gneisses are generally light in colour and contain many dark fragments of contorted schists. On Doctor lake the strike is north-and-south with a dip to the west. The Sturgeon Stone is a steep cliff over 100 feet high at the entrance to Doctor lake. Several small islands of sand and clay are seen in the centre of the channel and on the sides of the valley traces of a terrace still remain.

Above this the stream was not examined, but its course is from the west and is said to be very rough with many rapids and falls as far as the mouth of Deer river.

The hilly country to the south of Sisipuk lake and north-west of Loon lake is not well timbered but the lower land between the two and on the islands is fairly well covered by groves of small spruce.

On our trip to Nelson House we descended the Churchill to Nelson lake and went east up a long inlet or branch of Nelson lake to near the Burntwood river, to which we portaged. From Shaving point the river for a distance is flowing in a narrow crooked channel in which the current is about two miles an hour. This soon lessens as the channel broadens out to lake-like dimensions in which are scattered many large islands. The rocks are generally light-gray and whitish gneisses with fragments of dark schist and gneiss held as contorted inclusions. One long string of schist was observed to have been entirely folded back on itself. All these are again cut by flesh-coloured granite in large dykes. The general strike of the rocks seems to be about north-west and south-east. The hills here again become prominent and there are still on some of the islands and in sheltered bays traces of a terrace deposit of sand and clay. A short portage was made across a narrow neck at the centre of a long irregular island stretched across the course of the river. This road was cut through a small poplar grove growing on one of these terraces of clay.

Route taken to Nelson House.

Nelson House, formerly located on this lake, was an important post when the trade for the northern interior passed up the Churchill river. The site was pointed out on a small island at the south end of the lake opposite the channel leading to the portage to Burntwood river. At the present time the Hudson's Bay Company have established an outpost for winter trade on an island just within the narrow strait and therefore not very far from the old post. The rocks on the arm running to the east are light-coloured granites and gray and dark-coloured gneisses striking south-east and north-west. Occasional glimpses are had of the bare hills running along to the south. Several irregular bays branch off from the south side and by making two portages across narrow necks a straight course is formed which is much shorter and is generally followed by canoe parties. Two miles east from the present post the rock is a mass of granite dykes inclosing irregular areas and fragments of dark gneiss striking in several directions. At the first portage the rocks are dark garnetiferous gneisses striking south-east and north-west. The second portage is over a clay ridge filling up intervals between rocky ridges of light-coloured granite and gneiss. Both portages are of moderate length—the first 600 yards and the second 200 yards.

Nelson House.

The portage to Burntwood river starts from a small stream which flows from the east and enters the southern end of the long lake noted above through a gap in the ridge which bounds its southern shore. It begins in a willow swamp but soon gains a rocky ridge at half a mile

Portage to Burntwood river.

and again dips through a swamp. Half a mile farther a higher ridge is gained and from there on to the river there is a long clay slope or terrace covered by Banksian pine. At the river bank the vegetation is richer and in the tall grass is found the wild pea vine. Poplar trees replace the pine. It seems probable that the better drainage of the river bank makes the soil warmer and so encourages earlier growth